AMAZING SC

SABER-TOOTHED CATS

AND OTHER

PREHISTORIC WONDERS

Q.L. PEARCE

Illustrated by Mary Ann Fraser

Julian Messner

To my dad, with thanks for everything   —*Q.L. Pearce*
To Deborah, Robin, Ruth, Kathy, and Muriel —*Mary Ann Fraser*

Acknowledgments

With thanks to Grant E. Meyer, Director of the Raymond M. Alf Museum, Claremont, California, for his invaluable assistance and critical review of the manuscript. Thanks to J.D. Stewart of the Natural History Museum of Los Angeles County.

Library of Congress Cataloging-in-Publication Data
Pearce, Q.L. (Querida Lee)
  Saber-toothed cats and other prehistoric wonders / Q.L. Pearce;
illustrated by Mary Ann Fraser.
    p. cm.—(Amazing science)
  Includes bibliographical references.
  Includes index.
    Summary: Describes a variety of animals that lived long ago and their evolution into today's species.
  1. Animals, Fossil—Juvenile literature. 2. Evolution—Juvenile literature.
[1. Prehistoric animals. 2. Evolution.] I. Fraser, Mary Ann, ill. II. Title.
III. Series. Pearce, Q.L. (Querida Lee)
Amazing science.
QE675.P43 1991   90-20321
560—dc20   CIP
                                                                                 AC
ISBN 0-671-70691-8 (lib. bdg.)
ISBN 0-671-70692-6 (pbk.)

Text copyright © 1991 by RGA Publishing Group, Inc.
Illustrations copyright © 1991 by RGA Publishing Group, Inc.
All rights reserved including the right of
reproduction in whole or in part in any form.
Published by Julian Messner, a division of
Silver Burdett Press, Inc., Simon & Schuster, Inc.
Prentice Hall Bldg., Englewood Cliffs, NJ 07632.

JULIAN MESSNER and colophon are trademarks of
Simon & Schuster, Inc.
Manufactured in the United States of America.

Lib. ed.:   10 9 8 7 6 5 4 3 2 1
Paper ed.: 10 9 8 7 6 5 4 3 2 1

Contents

An Island of Life	5
Geologic Calendars and Radioactive Clocks	7
The Trilobites	9
Rebirth of an Ocean	11
Life in the Sea	13
The Age of Fish	15
Pioneer Plants	17
Creeping Ashore	19
The Great Frontier	21
Fantastic Forests	23
A New Breed	25
Mountain Building	27
Giant Birds	29
The Age of Mammals	31
The Birth of Mammals	33
Return to the Sea	35
The Great Cats	37
A Human's Best Friend	39
Megatherium and *Glyptodon*	41
Hoofs and Horns	43
Ancient Elephants	45
Horses: From Forests to Grasslands	47
The Largest Land Mammal	49
The Giant Deer	51
Earth's Changing Face	53
Gigantopithecus: The Giant Ape	55
Lucy: Our Distant Ancestor?	57
The First Tool-Makers	59
Ancient Art	61
Lessons from the Past	62
For Further Reading	63
Index	64

An Island of Life

Have you ever heard the saying, "There's no place like home"? There certainly is no place that we know of that's like our amazing home planet. Earth is the only island of life in the solar system. Hundreds of thousands of species of plants and animals live on the earth today, but they are only a fraction of the number that have ever lived. Over millions of years, some astounding creatures have inhabited this planet. Scorpions up to seven feet long hunted in the ancient seas, a giant bird the length of a modern pony once soared through the air, and huge shaggy-coated ancestors of the elephant lumbered over the prehistoric plains.

Earth, however, has not always supported life. About 4½ billion years ago, the earth was a spinning ball of molten rock. As it cooled, it formed a hard surface crust. Towering volcanoes belched poisonous gases into the early atmosphere. Storms raged and lightning struck the lifeless planet again and again. Water vapor in the atmosphere condensed and fell as rain. For tens of centuries, the rains poured down continuously, creating vast, shallow oceans. Violent waves smashed the barren shores. Yet, suspended in the churning, lifeless sea were the very elements that would form the first living things.

Roughly 3.5 to 4 billion years ago, chemicals in the water began to build new substances that combined in various ways to make the first living things. These were microscopic bacteria that "fed" on the chemicals in the water. The next great step was the appearance of cynobacteria (si-no-bak-TEER-ee-uh), which made their own food through a process known as photosynthesis. The by-product of the process was oxygen. As the cynobacteria prospered, they released more and more oxygen into the atmosphere. The stage was thus set for life to flourish in the sea, on land, and in the air. This book introduces you to some of the amazing creatures that inhabited the prehistoric world.

ERA	MILLIONS OF YEARS AGO	PERIOD	ANIMAL LIFE
CENOZOIC	5	QUATERNARY Recent Pleistocene	
CENOZOIC	65	TERTIARY Pliocene Miocene Oligocene Eocene Paleocene	
MESOZOIC	144	CRETACEOUS	
MESOZOIC	213	JURASSIC	
MESOZOIC	248	TRIASSIC	
PALEOZOIC	286	PERMIAN	
PALEOZOIC	320	PENNSYLVANIAN	
PALEOZOIC	360	MISSISSIPIAN	
PALEOZOIC	408	DEVONIAN	
PALEOZOIC	438	SILURIAN	
PALEOZOIC	505	ORDOVICIAN	
PALEOZOIC	590	CAMBRIAN	
PRECAMBRIAN		PRECAMBRIAN	

Geologic Calendars and Radioactive Clocks

We can think of the history of the earth as being similar to a book about "geologic time." Scientists have divided this "book" into "chapters" called eras. The first chapter, the Precambrian Era, takes up more than 90 percent of the book. Other chapters are the Paleozoic ("ancient life") Era, the Mesozoic ("middle life") Era, and the Cenozoic ("recent life") Era. Each chapter is divided into sections called periods. When a new fossil is discovered, how can we find out in which chapter or section of Earth's "book" it belongs?

One way to figure out the age of a fossil or a rock is to observe its location. Rocks formed from mud, sand, or clay are laid down in layers called strata. The oldest layers are on the bottom. The depth and thickness of strata are clues to the age of fossils discovered in them.

Certain rocks have "radioactive clocks" that help us to tell their ages. Everything in the universe is made up of tiny particles called atoms. Many kinds of atoms decay, or lose particles from their centers. These atoms are called radioactive. As they decay, they change into other substances. For example, over millions of years, radioactive uranium atoms decay into nonradioactive atoms of lead and helium. By comparing the amount of uranium to lead in a rock, scientists can determine how old the rock is.

There is also a way of determining the age of once-living things that are less than 100,000 years old. Living things take in a substance called carbon 14 from the air they breathe and the food they eat. As long as the plant or animal is alive, the amount of carbon 14 in its cells remains the same. When the living thing dies, however, the carbon 14 begins to decay. By measuring how much carbon 14 is left, we can estimate how much time has passed since the plant or animal died.

The "geologic book" that has recorded the history of life on Earth is divided into four eras.

The Trilobites

Science is often helped by chance. A minor accident in western Canada led to the discovery of a treasure trove of fossils. One day in 1909, a geologist named Charles Walcott was horseback riding on a mountain trail. Suddenly his horse stumbled. When Walcott dismounted to check its leg, he noticed an unusual rock. Recognizing it as shale, a rock formed from sea floor mud, he picked it up and cracked it open. Inside was a fossil, the first of thousands of fossils that would come from a huge deposit now known as the Burgess Shale. This area, which is high in the Rocky Mountains, was once 300 feet deep in an ancient sea! Scientists have learned that the fossils date from the dawn of the Paleozoic (pail-ee-o-ZO-ik) Era.

Since Walcott's discovery, more than 120 kinds of creatures have been found in the Burgess Shale. Among them are many odd sea animals called trilobites. Trilobites were among the first living things to develop hard outer shells. These shells were made of material similar to your fingernails. Trilobites belonged to the group of animals called arthropods. Modern arthropods include spiders, crabs, and insects. The warm, shallow seas of long ago must have been filled with trilobites. Thousands of different types have been discovered, ranging in length from less than one-tenth of an inch to more than two feet.

Some of the trilobite's early neighbors under the sea are also preserved in the Burgess Shale. *Opabinia* (o-puh-BEEN-e-uh) was a bottom dweller with five eyes. It had a long tube in front that looked something like a vacuum cleaner hose with fangs. Another odd animal was named *Hallucigenia* (huh-loo-suh-GEEN-ee-uh), which means "something out of a dream." The dream must have been pretty strange. This creature had seven pairs of stiff, toothpicklike legs, and seven tentacles along its back!

Thousands of different kinds of trilobites populated ancient seas, and with them, many strange neighbors.

Rebirth of an Ocean

The story of earliest life takes place in oceans and shallow seas. Like life itself, these watery worlds and their shorelines are always changing. Today, to travel from North America to Europe you must cross the Atlantic Ocean. If you had lived millions of years ago, you could have made the same journey in one step because the Atlantic Ocean did not exist! Earth's continents are passengers on great crustal plates that split, move apart, and rejoin at an extremely slow pace. Shallow seas filled the early splits, or rifts, and grew into mighty oceans as the continents drifted further apart.

During the past billion years, the body of water we call the Atlantic Ocean has opened and closed several times. During the Cambrian Period of the Paleozoic Era, an ocean we call Iapetus (eye-AP-uh-tus) separated North America, Europe, and Africa. Some scientists believe Iapetus was at least 1,000 miles wide. Others think it was a much smaller sea, dotted with many large islands. About 300 MYA (million years ago), the moving continents closed Iapetus and joined with the rest of Earth's land mass to become a supercontinent called Pangaea (pan-JEE-uh). The collision of these continents twisted and crumpled the ancient shorelines and raised towering mountain peaks. But Pangaea did not last. Great rifts violently broke it apart.

The modern Atlantic Ocean was born about 180 MYA. Once again, rocked by earthquakes, the land split and began to move apart. The seas slowly crept into the spreading gap. Today, the Atlantic is still widening. If current theories are correct, this mighty ocean will eventually close again, and a new supercontinent will emerge. But there is no reason to get rid of your world map yet. The breakup and reassembly of a supercontinent takes about 440 million years!

During the Cambrian Period, a narrow ocean called Iapetus separated North America from Europe and Africa.

Nautiloid

Sea scorp

Life in the Sea

Have you ever noticed that some days seem to go by very quickly? Well, days went by even faster during the Paleozoic Era. Because the earth was spinning on its axis more rapidly then, there were over 400 days in a year, compared to our present 365! This surprising fact is "recorded" in banks of coral reefs that grew in the ancient seas. Coral grows more during the night than during the day. By observing the bands of growth in prehistoric coral, we can tell the length of an average day at that time.

About 500 MYA, during the Ordovician (ord-uh-VISH-un) Period, the shallow seas were filled with coral. Tropical bays and lagoons fringed by coral reefs covered much of the northeastern United States. Sea creatures swam or crawled over what is now New York and Ohio. Trilobites, sea snails, starfish, prickly little sea urchins, and jellyfish were plentiful.

Another common sea creature was the squidlike nautiloid (NAW-tih-loyd). It moved by a sort of jet propulsion, sucking water in and squirting it out through a special opening. Tucked safely into the open end of its shell, the nautiloid cruised through the water backward, trailing snaky tentacles lined with sucker disks. The tentacles were perfect for snaring unwary prey. From tip to tip, this ancient "sea monster" could measure nearly 15 feet long.

The sea bottom is not a place where you might expect to find scorpions. However, during the Silurian (sy-LOOR-ee-an) Period, fierce sea scorpions ruled a watery domain. They had eight legs, segmented bodies, and sharp, slender tails. Some were equipped with flattened paddles at the ends of one pair of legs. Others had powerful pincers. The largest sea scorpion was the seven-foot-long *Pterygotus* (tair-ih-GO-tus). A ferocious hunter, *Pterygotus* probably preyed on anything it could catch, including members of a remarkable new class of animals: the fish.

The squidlike nautiloid and dangerous sea scorpion feasted on such smaller ocean creatures as trilobites and early fish.

The Age of Fish

Think of a fish. Do you picture an animal with fins? That's true of modern fish but not of their earliest ancestors. Six-inch-long *Arandaspis* (air-an-DAS-pis) was among the first fish, appearing during the Ordovician Period. It had a paddlelike tail, but no fins to steady it in the water. It also lacked jaws in its tiny mouth. A bony shield over its head gave *Arandaspis* some protection against predators. *Arandaspis* also had something special that set it apart from other animals: a primitive sort of "backbone" in the form of a stiff cord of cartilage.

Over millions of years, the fish that followed *Arandaspis* developed fins and became swift, agile swimmers. Many had overlapping bony scales that protected them from predators. By the close of the Silurian Period, a dramatic change had taken place in the fish. Stiff gill supports of cartilage had folded forward to form jaws. These "new" fish were called placoderms (PLAK-o-dermz), which means "flat-plated skins." Among their ranks was one of the most ferocious creatures in the sea, *Dinichthys* (dy-NIK-thus). This 30-foot-long predatory beast had a gigantic, armored head. A two-foot-long wall of sharp, jagged-edged bone lined its jaws. But *Dinichthys* wasn't the only predator at that time. It had to compete with early sharks, such as *Cladoselache* (klah-doh-sel-AK-ee).

The Devonian (deh-VOWN-ee-an) Period is called the "age of fish" because so many new varieties of fish appeared during that time. These include the bony fish, whose skeletons were made of bone. The two main groups of bony fish that evolved during the Devonian Period were the ray-finned fish, which gave rise to most modern types of fish, and the lobe-finned fish, which possibly had another destiny. A member of this group may have been the first backboned animal to struggle out of the sea and onto dry land.

A ferocious Devonian sea creature, Dinichthys *measured 30 feet long and had jagged-edged bone for teeth.*

Pioneer Plants

Can you imagine the world without plants? Without plants there would be no animals, because all animals depend directly or indirectly on plants for food. Earth would be a bleak and barren place. On land, the only sounds would be the howling of the wind and the crashing of waves on rocky shores. Until around 410 MYA, although plants and animals flourished in the sea, Earth's continents were empty. The first pioneering plants to venture out of the safety of the waters onto land were algae (AL-jee), living in tidal zones at the edge of the sea. At low tide, the algae were exposed to the air until the waters returned. Slowly, they adapted to the drier periods. Over millions of years, tiny plants crowded the shore to take advantage of the sunlight and carbon dioxide needed for photosynthesis.

Still, it was a giant step from shoreline to dry land. In the water, plants simply floated. They did not need roots because the water carried mineral nutrients directly to the plant. And, of course, water plants were seldom in danger of drying out. But on land, plants needed an entirely different design. Land plants needed roots to anchor them and to gather water and nutrients from the soil. A system of tiny tubes in a sturdy stem was required to carry the water and nutrients to the rest of the plant. Plants that followed this design are called vascular plants. The earliest known vascular plant was the two-inch-tall *Cooksonia*. As of 395 MYA, the ancestors of most of the plants we see today had developed.

Forests of miniature plants spread inland along riverbanks and lake shores. Plants had conquered a vast new world, but they weren't alone for long. Lured by the wealth of plant food, living space, and safety from the dangers of the seas, animal colonists soon left the water too.

About 400 MYA, plants took an enormous step—from the safety of the ocean to the new world of dry land.

Creeping Ashore

A millipede is a wormlike creature that holds some amazing records. Its name means "1,000 feet." That's an exaggeration, but the millipede does have up to 400 feet, or 200 pairs of legs—more than any animal that has ever lived. During the Devonian Period, the remarkable millipede may also have been one of the first animals to set foot (or feet!) on land. (These early pioneers also include insects, scorpions, and spiders.) The millipede probably nibbled its way out of the sea through soaking mats of algae that carpeted the shoreline. Because it had a tough outer skeleton, the millipede was able to crawl along without the support of water. Still, it had to adapt to breathing in air instead of water before it could be truly free of the sea. The first ancient millipedes were probably small, but by the Carboniferous (kar-bo-NIF-er-us) Period, some had reached the astounding length of more than six feet. That's about as long as a horse!

It was probably not very long before the dance of predator and prey began on land. Meat-eating scorpions and early spiders crept out of the sea to feed on the plant-eaters. Scientists have discovered a fossilized portion of a silk-spinning spider that is more than 380 million years old. They have also found ancient fossil leaves that appear to have been munched on by hungry insects.

The first insects were tiny, wingless varieties known as bristletails and springtails. The silverfish, a modern bristletail, looks very much like its ancestors did. From those small beginnings, insects have developed to become the most successful form of animal life on Earth. For every human being alive, there are about one million insects!

Among the first to creep ashore: the plant-eating millipede, the meat-eating scorpion, and the tiny bristletail insect.

Diplocaulus

The Great F[...]

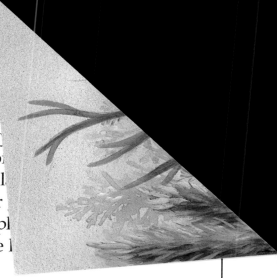

At the beginning of the Devonian P[...] populated and dry land was a great [...] boned animal to wriggle ashore ma[...] finned fish escaping from a predato[...] too difficult for this fish to adapt to l[...] already equipped with an air sac for [...] face when oxygen was in short suppl[...] had muscular fins with bones in the l[...] on the muddy sea bottom.

By the end of the Devonian, some of the fins of the first pioneer fish had become feet. These fish had developed into amphibians (animals that live in both water and on land). The earliest known amphibian was three-foot-long *Ichthyostega* (ik-thee-o-STEG-uh), whose name means "fishlike plates." It had sturdy legs and a short neck so that it could turn its head from side to side. Still, *Ichthyostega* had a tail fin and fishlike scales, and it was probably more comfortable in the water than on land.

Although amphibians never truly conquered the land, over a span of about 160 million years they developed useful traits such as a simple eardrum to detect airborne sounds, and eyelids and tear glands to protect their eyes and keep them moist. *Platyhystrix* (plat-ee-HISS-triks) developed a tall, spiny "sail" on its back. The first known plant-eating amphibian, *Diadectes* (dy-uh-DEK-teez), was very stocky and about 10 feet long. One of the oddest looking amphibians was swamp-dwelling *Diplocaulus* (dip-lo-KAWL-us). It had backswept bones on each side of its flat head. Maybe these bones made *Diplocaulus* hard for predators to swallow! By the beginning of the Jurassic (joor-ASS-ik) Period in the Mesozoic (mez-o-ZO-ik) Era, most of the early amphibians had become extinct. However, they had given rise to the families that include modern toads, frogs, newts, and salamanders.

■ Ichthyostega *was the earliest known amphibian and* Diplocaulus *was perhaps the strangest looking.*

Meganeura

Modern dragonfly

Fantastic Forests

Centuries ago, people thought coal grew from seeds in the ground because they often found the imprints of leaves inside chunks of coal. Actually, they were on the right track. Although it does not grow from seeds, under the proper conditions, coal forms from ancient plant life. The Carboniferous Period of the Paleozoic Era is so named for the bands of coal (a form of carbon) that developed during that time. The bands formed from immense forests that shrouded swampy lowlands. When a tree died, it plunged into the stagnant water and sank beneath the surface before it could rot. Trees fell for millions of years, and each new layer pressed down on the soggy layers below. Eventually, heat and pressure turned the layers of trees into coal.

If you could travel through the ancient forests that gave birth to today's coal bands, or seams, they would appear strange and alien. Gigantic tree ferns towered above the mucky swamps. While modern mosses are no more than a few inches tall, in the steamy Carboniferous forests, club mosses such as *Lepidodendron* (leh-PID-o-DEN-drun) soared to more than 100 feet tall! At the foot of these giants, seed ferns such as *Medullosa* (med-yoo-LOW-suh) grew in luxurious green clusters 15 feet high.

In the eerie shadows below, amphibians hunted for fish in the water or lurked on muddy banks in search of insects. Cockroaches and beetles the size of your hand scurried among tangles of roots and leaf litter. Insects also probably climbed high into the trees to search for food and to escape hungry amphibians. Also during this period the first flying insects entered the fossil record. One of these was the dragonfly *Meganeura* (meg-uh-NOOR-uh), the largest flying insect that has ever lived. With a wingspan of more than two feet, this dragonfly was about the size of a small crow.

Forests of gigantic tree ferns and eerie mosses were home to Meganeura, *the largest flying insect that has ever lived.*

Cynognathus

A New Breed

Did you know that if you traced the mammal family tree to its beginnings you would find a class of reptiles at its roots? As the Paleozoic Era was drawing to a close, the climate became cooler and drier. The great swamps were replaced by dry forests of pine and fir, and arid deserts stretched far inland. Small, early reptiles, with their scaly, watertight skin and eggs with shells, prospered in this new world. One of the four major groups of reptiles was the synapsids (sih-NAP-sidz), or mammal-like reptiles.

Dimetrodon (dih-MEE-tro-don) was a pelycosaur (peh-LEE-ko-sor), a very early form of synapsid. This 9-foot-long, 400-pound beast bore little resemblance to any modern mammal. Like other reptiles, *Dimetrodon* was cold-blooded, but it had a "sail" of leathery skin on its back that helped to regulate its body temperature. This sail was supported by spines that were more than three feet long at the center. By turning the sail to catch the rays of the rising sun, this fierce predator could warm itself twice as fast as other reptiles that it preyed upon, thus gaining an advantage. And unlike most other reptiles, *Dimetrodon*'s teeth were not all alike. Its powerful jaws were lined with biting and grasping teeth, as well as with stabbing canine teeth.

Although it was a reptile, *Cynognathus* (sy-no-NAY-thus) looked more like a mammal and was most likely warm-blooded. A member of a group of synapsids called the therapsids (ther-AP-sidz), the fast *Cynognathus* was a wolf-sized meat-eater. Its teeth and jaws were similar to those of a dog, and it had fleshy, external ear flaps. In fact, it may even have had a covering of fur.

Although the synapsids seemed to be off to a good start, they were not destined to conquer their world. Instead, a different group of reptiles went on to dominate the land for nearly 160 million years: the diapsids (dy-AP-sidz), a group that included the dinosaurs.

The reptile Dimetrodon *had mammal-like qualities, as did furry* Cynognathus, *a wolf-sized meat-eater.*

Mountain Building

Mount Everest, the highest mountain on Earth, towers 5½ miles above sea level. It is part of a tremendous 2,000-mile-long range of mountains called the Himalayas. Covering an area the size of Germany, these mountains form a natural border between the continent of Asia and the subcontinent of India. Besides Mount Everest, the Himalayas include 14 other majestic peaks that sweep upward for more than four miles each. And the range is still rising! But part of this "roof of the world" was once at the bottom of an ancient sea bed. Scientists have discovered impressions of fossil marine animals in rocks 15,000 feet above sea level and 300 miles from the nearest modern ocean.

Mountains form in several ways. The mighty Himalayas are "fold" mountains, which means they were thrust skyward when two of Earth's crustal plates slowly crunched together. Soon after the supercontinent Pangaea began to break apart, India separated from Africa and Antarctica. It began to drift on a 4,000-mile journey toward Asia, moving at a rate of about 15 to 30 feet per century. As it approached Asia, the heavy, deep ocean floor between the two land masses was pushed under the lighter rocks of the Asian continent and down into Earth's molten mantle. Eventually, only a narrow, shallow sea separated the two continents. India continued to drive northward and, about 40 MYA, it collided with Asia. The continental rocks of India didn't follow the deep ocean floor down into the mantle. Instead, they began to buckle upward. Over millions of years, the land squeezed together, rose, and toppled into huge folds. In the process, remnants of the shallow sea were lifted up and stranded high on the craggy Himalayan mountain peaks. But these great mountains are not here to stay. Millions of years from now, the forces of erosion will carry the peaks of the Himalayas back into the sea, bit by bit.

The mighty Himalayas have formed as a result of the Indian plate crushing against the Asian plate for millions of years.

Giant Birds

Maybe you've heard this joke: "What is yellow and weighs 1,000 pounds?" The answer: "Two 500-pound canaries." Of course, no bird could weigh that much . . . or could it? Actually, the class of animals that makes up the birds has included a number of creatures even larger than this.

A family of large birds with stout, hooked beaks and sharp, sturdy claws lived at the beginning of the Cenozoic (sen-o-ZO-ik) Era. These birds were called the terror cranes. The largest of them was *Diatryma gigantea* (dy-uh-TRY-muh jy-GAN-tee-uh), which stood seven feet tall. Although its stubby wings were useless for flight, this robust bird could move quickly over the ground on its powerful, scaly legs. Some scientists think that *Diatryma* preyed on small mammals, but others think the terror bird was a harmless plant-eater that used its sharp-edged beak to clip vegetation.

Aepyornis (ay-pee-OR-nis) was a giant slow-moving bird that weighed about 1,000 pounds—four times as much as a newborn elephant. Commonly called the elephant bird, *Aepyornis* belonged to a large group of flightless birds known as ratites. (The modern ostrich belongs to this group, too.) It also laid the largest eggs ever. A single *Aepyornis* egg could hold 12,000 hummingbird eggs! *Aepyornis* lived in Madagascar during the Pleistocene (PLYSS-toh-seen) Epoch of the Quaternary (KWAT-er-nair-ee) Period and beyond—until just three centuries ago, when it was hunted into extinction by humans.

The largest flying bird ever, *Argentavis magnificens* (ar-gen-TAY-vis mag-NIF-ih-senz), was an early vulture of the Tertiary (TER-shee-air-ee) Period. *Argentavis* was a scavenger. As long as a pony and weighing as much as a mountain lion, it glided over the grasslands of South America searching for carcasses to feed on. Its wingspan was 24 feet—the largest of any bird that has ever lived.

■ *The largest terror crane was the flightless* Diatryma gigantea, *which could move quickly over ground in search of food.*

The Age of Mammals

Can you imagine what it would be like to live in a world of giants, where other creatures were 10, 20 . . . even 100 times bigger than you? The world must have been something like that for the earliest mammals. Often small enough to have fit in your hand, these mammals lived during the age of dinosaurs. But then, at the close of the Mesozoic Era, something ended the dinosaurs' reign, yet spared the tiny mammals. One theory is that a huge comet or asteroid collided with Earth, raising a choking blanket of dust and ash that circled the planet. Several dark, cold months passed before the atmosphere cleared. Without sunlight, green plants were unable to produce their own food and died. Without plants, plant-eating dinosaurs soon starved to death. They were followed by the great flesh-eating dinosaurs. When the dinosaurs were gone, the mammals crawled from hiding to claim the earth, and the Cenozoic Era—the age of mammals—began.

But how did these small creatures survive the catastrophe that claimed the mighty dinosaurs? The mammals were night creatures, well-equipped to get by in the new world. They were also warm-blooded animals, which means that their body temperatures stayed fairly constant, no matter what the temperature of the outside environment. A coat of fur helped to protect them from the cold. (Some may have hibernated and so avoided the problem of cold completely.) It's likely that many of the mammals were omnivorous—that is, they were able to eat whatever food was available. *Purgatorius* (pur-guh-TOR-ee-us) was such a creature. Only four inches long and weighing about as much as a sparrow, it probably survived by eating burrowing insects and worms, insect pupae, and seeds. Its endurance is fortunate for us, because this tiny animal may have been the ancestor of all primates, a group that includes monkeys, apes, and humans.

Tiny Purgatorius, *one of the earliest mammals, survived by eating insects, worms, insect pupae, and seeds.*

The Birth of Mammals

Did you know that some mammals lay eggs? The duck-billed platypus and the echidnas of Australia do. Known as monotremes, they are the only surviving members of the earliest mammal group, the prototheria. By observing modern monotremes, we can guess what the other prototheria were like. The tiny, shrewlike *Megazostrodon* (meg-uh-ZOS-tro-don) female may have laid eggs, too. She probably cared for her hatched young by feeding them milk from her body.

By the Cretaceous (crih-TAY-shus) Period, a new group of mammals had developed. This group was the theria, which includes the metatheria (marsupials, or pouched mammals) and the eutheria (placental mammals). The difference between pouched and placental mammals is in how the babies are born. In pouched mammals, such as modern kangaroos, the young develop partly in the mother. They then leave her body and crawl into her belly pouch until they develop enough to be on their own. *Diprotodon* (di-PROT-o-don) was the largest marsupial ever. It was 11 feet long and looked a little like a furry hippo.

Placental mammals make up more than 90 percent of all mammals, past and present. Their babies are well developed at birth and grow quickly. Creodonts were flesh-eating members of this group. Among their ranks was ferocious *Megistotherium* (meh-GIST-o-THEER-ee-um), one of the largest flesh-eating mammals on Earth. About 10 feet long, this beast weighed nearly 2,000 pounds. It had a massive head, twice as large as a modern bear's, and sharp, heavy teeth. For millions of years, *Megistotherium* and other mammals like it were the mightiest predators on the planet. But as time passed, their prey became faster and more agile. The creodonts gave way to the carnivores, who were more capable hunters. Carnivores today include lions, tigers, and wolves.

One of the mightiest of predators during the Cretaceous, Megistotherium used its massive teeth to snare a victim.

Return to the Sea

If you were to close your eyes and point to any spot on a globe or map, you would most likely be pointing to a place where mammals live. Whales and dolphins are mammals that roam throughout the world's oceans. They are adapted to life in the water, but a little less than 60 MYA their ancestors were still plodding about on land. *Pakicetus* (pak-ee-SEE-tus), the earliest known whale, lived during the Eocene (EE-o-seen) Epoch. Only the remains of a *Pakicetus* skull have been found, but scientists speculate that this creature had stout, paddle-shaped legs and may have dined on fish. Still, *Pakicetus* wasn't suited to life in the water. It couldn't see or hear well below the surface, and it couldn't dive very deep. This creature probably settled near the shore, spending much of its time on land.

Basilosaurus (baz-il-o-SOR-us) was a 75-foot-long, ocean-dwelling whale with a slender, legless body and a mouth full of saw-edged teeth. Powered by fins and tail flukes, *Basilosaurus* chased schools of tasty fish or wriggly squid, even in deep water. Like all mammals, it had to breathe air. Its nostrils were high on its snout, making it easier for the animal to catch a breath at the surface. The female *Basilosaurus* probably even gave birth at sea.

Another group of sea mammals, the pinnipeds, or "fin-footed" mammals, evolved about 30 MYA. Pinnipeds today include seals, sea lions, and walruses. *Enaliarctos* (ee-NAY-lee-ark-tos) was a sleek shore creature of the Miocene (MY-o-seen) Epoch. This meat-eater had short, stout legs and was probably a clumsy hunter on land. But its long slender toes were webbed, so it was able to cut gracefully through shallow waters to gather a meal of fish or shellfish. Like its modern relative the sea lion, *Enaliarctos* had large eyes and an excellent sense of hearing, even underwater. Perhaps this ancient sea mammal was as playful as its descendants are today.

■ *Like its modern sea mammal relatives, 75-foot-long* Basilosaurus *breathed in air through nostrils on the top of its head.*

Smilodon skull Modern lion skull

The Great Cats

You've probably heard of state birds and flowers, but have you ever heard of a state fossil? The saber-toothed cat called *Smilodon* (SMY-lo-don) is the state fossil of California. Hundreds of these predators were entombed in tar pits in an area that is now part of Los Angeles, called Rancho La Brea. For millions of years, sticky tar has oozed to the surface in pools topped by rainwater. Animals that waded into the pools to drink were trapped in the gooey tar. Their cries brought hungry meat-eaters, such as *Smilodon*, to claim what these predators thought would be an easy meal. But the great cats, too, met certain death as they became caught in the tar. The number of saber-toothed cats that perished in the tar at Rancho La Brea indicates that a great many of these beasts roamed the American Southwest during the Pleistocene Epoch.

Weighing more than 600 pounds, *Smilodon* was the size of a large lion of today. But this remarkable animal was equipped with weapons unlike those of any modern cat. It had two daggerlike teeth in its upper jaw, each of which was up to eight inches long. In fact, the name *Smilodon* means "knife-tooth." Like the modern lion, this cat stalked its prey and drew as close as possible before it pounced. Leaping on a thick-skinned creature, such as a giant ground sloth or a bison, *Smilodon* held on with its razor-sharp claws. Dropping its lower jaw out of the way, the great cat stabbed its fangs deep into the animal's flesh, and the prey animal simply bled to death from the terrible wounds.

The largest of all prehistoric cats was the cave lion of Europe. Nearly 12 feet long, the cave lion was about one-third larger than a modern lion. Unlike *Smilodon*, it was a biting cat with short, sharp canine teeth. No less deadly than its saber-toothed relative, the cave lion quickly broke the neck of its prey with one powerful bite.

With its great stabbing teeth, Smilodon *attacked such unfortunate trapped prey as this bison.*

Phlaocyon

A Human's Best Friend

Often called a human's best friend, the domestic dog and its wild relatives have a very long history. The canine family tree (which includes wolves, coyotes, jackals, and foxes) took root about 40 MYA. One of its earliest members looked rather like a chubby raccoon, much like the modern raccoon dog of Asia. About 2½ feet long, *Phlaocyon* (flay-o-SY-on) was better at climbing trees than at running long distances. On the ground, it probably waddled from place to place nosing out seeds, fruit, eggs, insects, and an occasional bird or rodent.

If you came face to face with the dire wolf of the Pleistocene Epoch, you would certainly recognize it as a member of the dog family. Although it was stockier, the dire wolf closely resembled a modern timber wolf. This remarkably successful predator had strong jaws with a battery of cutting, crushing, stabbing, and shearing teeth. It devoured almost anything, from meat and bones to insects and fruit. We can be fairly certain of the dire wolf's appearance because, like saber-toothed cats, dire wolves are well represented at the Rancho La Brea tar pits. In fact, these tar pits hold the remains of more dire wolves than of any other type of animal.

Dire wolves most likely had keen senses, particularly a superb sense of smell. We can only guess about their hunting tactics, but they may have hunted in packs, as modern wolves do. Although they may not have been any faster or smarter than their prey, dire wolves probably had greater endurance, allowing them to relentlessly pursue an animal until it was exhausted. However, because of the number of dire wolves in the tar pits, some scientists think that these animals were scavengers rather than hunters, and were attracted by the wails of animals trapped in the tar.

Phlaocyon was one of the first members of the dog family, as was the dire wolf, who probably hunted in packs for food.

Megatherium *and* Glyptodon

An old legend in Argentina tells of gigantic, bearlike animals that once lumbered across the land. Such creatures did roam the grasslands and open woodlands of South America long ago. They were ground sloths, extinct members of the mammal group that includes modern anteaters, tree sloths, and armadillos. The giant ground sloth of the Pleistocene Epoch, *Megatherium* (meg-uh-THEER-ee-um), or "great beast," was the grandest of them all.

At the shoulders, *Megatherium* was as tall as an adult human. From nose to tail it stretched 20 feet long. Large individuals weighed three tons—twice the weight of an African black rhinoceros of today. Although thick, curved claws tipped its sturdy limbs, this beast was a plant-eater. Rearing up on its hind legs to reach tall branches, *Megatherium* steadied itself on its thick tail, then snagged a leafy branch with its hooked claws and drew it close.

This giant ground sloth plodded along on the sides of its three-foot-long hind feet and on the knuckles of its forefeet (to protect its claws). This wasn't a very efficient way of getting around, and *Megatherium* was probably slow-moving. This made it an excellent target for its predatory neighbors, in spite of its great size.

Not all of *Megatherium*'s neighbors were ferocious, however. Ten-foot-long, five-foot-high *Glyptodon* (GLIP-toh-don) preferred to be left alone to munch peacefully on tall, tough grass. Unlike the ground sloth, *Glyptodon* had an excellent system of defense against predators. It carried a dome-shaped shell on its back, a bony shield on its head, and a tube of bony rings on its hefty tail. *Doedicurus* (doh-DIK-ur-us), a relative of *Glyptodon*, took this strategy one step further. Its tail ended in a cluster of sharp spikes that could be used as a mighty club.

Megatherium had only its claws for protection, while smaller Doedicurus was armed with a cluster of sharp spikes on its tail.

Arsinoitherium

Hoofs and Horns

Before setting out on a hunt, American Indians of the Great Plains would sometimes put their ears to the ground to locate a herd of bison. The ground is an excellent conductor of sound, and the heavy hoofbeats of the huge animals carried for great distances. Bison, caribou, and horses are just a few of the many animals called ungulates (UN-gyuh-layts), or hoofed mammals. Today, ungulates are plant-eaters, but this hasn't always been the case. Sheep-sized *Phenacodus* (feh-NAY-ko-dus) wandered woodland glades during the Eocene Epoch. It had both stabbing canine teeth for eating meat and grinding cheek teeth for munching on plants.

Some ungulates developed incredible headgear. *Uintatherium* (win-tuh-THEER-ee-um) had a bizarre collection of three pairs of knobby, skin-covered horns on its three-foot-long head. The horns may have been used in contests of strength between males. Males also had long, sharp canine teeth, or tusks. Among the largest animals of the Eocene Epoch, *Uintatherium* was 6 feet high at the shoulder and nearly 13 feet long. It had stout, sturdy legs to bear its tremendous weight, which could easily be more than two tons. In spite of its great size, this colossal plant-eater had a brain no larger than that of a modern dog.

Scientists are baffled by *Arsinoitherium* (ar-sin-o-eh-THEER-ee-um). This unusual creature doesn't fit neatly into any other group of animals that has lived before or after it. During the Oligocene (ALL-ih-go-seen) Epoch, this ungainly animal sloshed through swampland in what is now Egypt. Although *Arsinoitherium* was nearly 12 feet long and 6 feet high, it is not its size but its two cone-shaped, hollow horns that make it such an amazing creature.

Uintatherium had a bizarre collection of six knobby horns on its head, while Arsinoitherium *had two hollow, cone-shaped horns.*

Amebelodon

Ancient Elephants

The only elephants you'll see in the United States today are those in zoos and circuses. But a little more than 10,000 years ago, elephant ancestors such as mastodonts and mammoths wandered freely over much of the land. *Amebelodon* (am-uh-BELL-o-don) belonged to an early group of North American mastodonts known as "shovel tuskers." It's easy to see how it got that name. Besides having a pair of short tusks in its upper jaw, this 10-foot-tall beast had two broad, bladelike teeth, each a yard long, protruding from its lower jaw. Plodding through shallow rivers, *Amebelodon* used these teeth to "shovel" up water plants that grew near the banks. It then used its flexible trunk to guide the plants into its mouth.

When mammoths ranged the earth, the northern lands were in the grip of the Ice Age. Throughout the Pleistocene Epoch, glaciers advanced and retreated. The land that is now southern England, Germany, and Poland was covered by cold grasslands. This was the home of *Mammuthus trogontherii* (mah-MOOTH-us tro-gon-THEER-ee-eye). At 15 feet tall, it was one of the largest mammoths. Its tusks were quite spectacular, curving from the upper jaw to a length of up to 17 feet.

The dark coat of *M. primigenius* (M. prim-ih-JEN-ee-us), better known as the woolly mammoth, was about 12 inches long and had a thick, woolly underlayer. A creature of the frozen tundra of Eurasia and North America, the woolly mammoth was well suited for cold weather. Besides its warm coat, it had a 3-inch-thick layer of blubber beneath its skin and could withstand temperatures to −50°F. When snow and ice coated the ground, this mammoth used its 10-foot-long tusks to scrape through to grass below. Food may often have been in short supply during the long winter, and scientists think this animal may have stored a fat reserve in a bulging hump behind its forehead!

Both Mammuthus primigenius *(the woolly mammoth) and* Amebelodon *had enormous, powerful tusks.*

Hyracotherium

Mesohippus

Merychippus

Pliohippus

Equ

Horses: From Forests to Grasslands

Did you know that the horses that pulled immigrants' wagons or plows during the settling of the American West were in a sense immigrants, too? Spaniards had introduced them to the Americas early in the 16th century. But it was not the first time that horses had roamed those plains. In fact, the first link in the horse line appeared in North America. Some 50 MYA, *Hyracotherium* (hy-RAK-o-theer-ee-um), also called *Eohippus* (ee-o-HIP-us), or "dawn horse," browsed on leafy undergrowth in Eocene forests. This intelligent animal was the size of a lamb and had four toes on its front feet and three on its back feet.

If there was a halfway point between the dawn horse and the modern horse, it may have been *Mesohippus* (mez-o-HIP-us), or "middle horse." Of the three toes on its front feet, the center toe was largest. *Mesohippus* still ate leaves and kept to the safety of the shadowy forest, but with its long legs, it could probably run at a good clip.

By the Miocene Epoch, many forested areas had given way to fertile grasslands that stretched for hundreds of miles. Over millions of years, the horse changed to survive in this habitat. To feed on the tough grass, donkey-sized *Merychippus* (mayr-ee-KIP-us) developed long grinding teeth and the long snout and deep jaw it needed to hold them. Although it still had three toes on each front foot, it ran only on the center one.

By 2 MYA, large herds of the modern horse *Equus* (EK-wuss) galloped over the vast grasslands. Then, about 8,000 years ago, most of the horses vanished. Disease is suspected, but their disappearance is still a mystery. Fortunately, some herds had spread to Asia by way of a land bridge that extended across the Bering Strait. Eventually, the horse returned to its original home—not by land this time, but by sea, on the ships of Spanish conquistadores.

Lamb-sized Hyracotherium, *which lived 50 MYA, was the first horse in the evolution toward the modern* Equus.

The Largest Land Mammal

Twice as tall as an elephant and as long as a school bus, *Indricotherium* (in-DRIK-o-theer-ee-um) was the largest land mammal ever to walk the earth. The tallest modern giraffe would have had to stretch its neck to look this beast in the eye! This stocky, 26-foot-long giant was an ancient relative of the rhinoceros. In fact, the largest living rhino, the great Indian rhinoceros, now plods through open forest in northern India not far from where *Indricotherium* lumbered along between 10 and 25 MYA.

Indricotherium was similar to its descendants in many ways. It had a slender, tufted tail, three hoofed toes on each foot, and broad, pillarlike legs to support its considerable bulk. But unlike modern rhinos, this massive creature weighed a whopping 30 tons or more—at least four times the weight of the Indian rhino.

The word *rhinoceros* means "nose horns" in Greek, but this early member of the family had none. Although its massive head was more than four feet long, it lacked a single horn. *Indricotherium* was a browser. With its long, muscular neck, it was able to reach branches that grew 25 feet above the ground. Its elongated, flexible upper lip efficiently gripped each branch and drew the tasty leaves toward the animal's mouth. *Indricotherium*'s odd front teeth—two forward-pointing teeth in the lower jaw and two tusklike teeth in the upper jaw—may have been useful for clipping tender twigs. Gathering in small herds, these ancient beasts browsed in dry, fairly open land. Although they had no apparent weapons for defense, their size alone probably offered them a great deal of protection.

■ *Indricotherium was a docile creature that dined only on leaves and twigs from the tallest trees.*

The Giant Deer

What do cattle, camels, pigs, and giraffes have in common? They all belong to the same large group artiodactyla (ar-tee-o-DAK-til-uh). This word means "even fingers" in the Greek language, and all members of this clan have either two or four toes on each foot. Included in this group are deer. Bony antlers generally crown the heads of male deer, from the simple antlers of the barking deer to the elaborate headgear of the reindeer. The largest member of the deer family today, the moose, may be 10 feet long and so tall that the top of an adult human's head would only reach the moose's shoulders. The enormous antlers of the moose are several feet wide. Even so, they cannot compare with the antlers of the magnificent *Megaloceros* (meg-uh-LAHS-er-us).

Often inappropriately called the Irish elk, *Megaloceros* ranged across much of Europe and Asia during the late Pleistocene Epoch. This incredible creature proudly displayed many-pronged antlers that spread some 12 feet or more from tip to tip. A particularly large pair of antlers easily weighed more than 100 pounds. To carry such a tremendous load, *Megaloceros* needed a strong, muscular neck. During the mating season, rival males, or stags, often staged ritual battles over females by wrestling each other with their massive antlers. Modern deer shed their antlers each year and grow a new pair to replace them, and there's no reason to believe that it was any different for *Megaloceros*.

The natural enemies of *Megaloceros* were probably large cats, wolves, and, eventually, humans. The ancient deer was the subject of many cave paintings, perhaps done by hunters to bring good luck in the hunt. The paintings show *Megaloceros* as having a full, dark coat of hair and an odd hump at its shoulders. The hump may have stored fat to sustain the animal through a harsh Ice Age winter.

The many-pronged antlers of Megaloceros, *which weighed more than 100 pounds, must have been an impressive sight.*

Earth's Changing Face

In 1961, researchers aboard a ship called the *Chain* were studying the contours of the floor of the Mediterranean Sea, a body of water 2,500 miles long and 850 miles across at its widest point. They discovered more than they expected. Far offshore, in deep water, lay salt domes of the sort generally found on dry land. How could they have formed at the bottom of a sea?

The answer lies in a story that began millions of years ago, when Africa split from South America and drifted northward. As Africa eased toward Europe, the Tethys (TETH-us) Sea, which separated the two continents, began to shrink. The surviving body of water, the Mediterranean Sea, was fed by several rivers, but mostly by the Atlantic Ocean through a narrow neck of water called the Straits of Gibraltar. As Africa continued to press on toward Europe, the land at Gibraltar was forced upward in the process. About 8 MYA, the rising land cut off the important connection between the Mediterranean and the Atlantic.

The waters of the ancient Mediterranean probably evaporated at a rate of about 1,000 cubic miles per year. A little of the water was replaced by rain and rivers, but the sea could not survive without being replenished by the Atlantic. In only a thousand years, there was nothing left but hot, salty pools and dry salt plains. In some places, the scorched valley reached temperatures of up to 150°F.

The powerful waters of the Atlantic could not be held back forever, though. About 5 MYA, the forces of erosion finally conquered the barrier between the Mediterranean and the Atlantic. Ocean water roared into the dry basin in the form of a magnificent waterfall. At least one mile high and carrying a volume of water a thousand times greater than that of Niagara Falls, the raging water poured in for more than 100 years. The desert was once again sea.

The Mediterranean Sea, once empty, was filled in 100 years by a waterfall created 5 MYA by the Atlantic Ocean.

Gigantopithecus: The Giant Ape

What sort of place do you think would be good for fossil hunting? Perhaps investigating a dry riverbed or the base of a seaside cliff would turn up a fossil or two. But would you consider a drugstore? In the 1930s, a scientist discovered several fossil teeth among the items for sale in a Hong Kong drugstore. It seems that the owner believed the teeth were quite special, and that when ground down, they would produce a powder with healing powers.

The teeth were special indeed. Twice the width of those of a modern gorilla, they had belonged to the largest member of the ape family that ever lived: *Gigantopithecus* (jy-gan-toh-PITH-ih-kus). From later discoveries of complete lower jaws, researchers calculated that this robust giant may have been as tall as a one-story building! It could easily have weighed 650 pounds or more. It probably looked much like a modern gorilla, but more on the scale of a junior King Kong. Its eyes, which peered out from under heavy brow ridges, were forward facing over a shortened snout. This allowed the animal to focus on an object with both eyes and so judge distance better.

Fur-covered *Gigantopithecus* was a ground-dweller. It ate mainly roots and seeds, and occasionally it also ate very small animals. It held its food with large, grasping hands tipped with nails instead of claws. *Gigantopithecus* usually plodded along on all four legs, keeping the fingers of its front limbs turned under so it could bear weight on its knuckles. We know little about the family life of this great ape, which once roamed the highlands of India, Pakistan, and China. It disappeared some 1 MYA and left few traces behind. But natives who live in the foothills of the Himalayas whisper stories of huge apelike creatures they call the Yeti, and so many people claim that *Gigantopithecus* may be with us still!

As tall as a one-story building, Gigantopithecus *roamed the highlands of the Far East, eating primarily roots and seeds.*

Fossil remains of Lucy

Lucy: Our Distant Ancestor?

Tracing the human line back to its beginnings isn't easy. Although humanlike creatures have lived for millions of years, the fossil trail is sparse. We don't know for certain which creature took the first step toward humankind, but a being known as *Australopithecus afarensis* (aws-tro-lo-PITH-eh-kus af-uh-REN-sis), meaning "southern ape from Afar," was on the right track. In 1974, the first fossil example of this creature, a female, was unearthed in northern Ethiopia in an inhospitable place called the Afar Triangle. As researchers excitedly examined the remains, a popular Beatles tune was playing on the radio—"Lucy in the Sky with Diamonds." So the scientists dubbed this likely distant ancestor of humans "Lucy."

About 20 years old when she died 3½ MYA, Lucy was neither ape nor human. Her brain was larger than that of an ape, but it was at least one-third smaller than that of a modern human. Unlike an ape, she had small canine teeth instead of fangs, and she walked upright without shuffling or stooping. We can't be certain, but she may have been dark-skinned, with a light covering of dark hair.

It's likely that Lucy, who was smaller than a modern eight-year-old child, lived with a small group that had banded together for care and protection. The group drifted from place to place, depending on the weather and the food supply. In bad weather they might have taken shelter in caves, but most often they lived in open woodland near rivers or lakes. These peaceful and cooperative beings probably shared such food as fruits and nuts, and perhaps also some fish, eggs, and meat. Still, life was harsh and unpredictable for Lucy and her kind. They died out around 2½ MYA—only a geologic moment before the first true humans began to walk the earth.

By piecing together the fossil remains of Lucy, scientists have an idea of what the earliest humans looked like.

Homo habilis tool kit		*Homo erectus* tool kit	
Scraper	Chopper	Cleaver	Hand axe
Blade	Flake	Pick	Cutting tool

The First Tool-Makers

Humans aren't the only creatures who use tools. A chimpanzee may use a slender twig to pry insects from a hole. A fishing otter often employs a rock to hammer open the shell of a shellfish. There's a difference, however, between using a natural object as a tool and actually making a tool.

The first creature that unquestionably appears to have made tools is also the first representative of the genus *Homo*, or "human." *Homo habilis* (ho-mo HAB-ih-lus), which means "handy human," appeared between 1½ and 2 MYA. Human teeth were not efficient at tearing through stringy meat or cracking hard nutshells, so *Homo habilis* used such sharp objects as shells or heavy rocks to do the job. With his large brain, excellent grip, and good hand/eye coordination, he was also able to fashion special tools. By chipping away at rocks, he created sharp-edged pebble tools. *Homo erectus* (ho-mo ee-REK-tus), or "upright man," was even more skillful. These humans developed stone hand axes for hunting.

As humans advanced, so did their tool-making ability. *Homo sapiens neandertalensis* (ho-mo SAY-pee-enz nee-an-der-tuh-LEN-sis), the Neandertal man, fashioned more than 60 different tools. Some cutting tools were saw-edged and others had special notches. Most were chipped from rock or carved from bone or antler. (Tools made of wood, skin, or fiber would, of course, have rotted away long ago, leaving no clues behind.) *Homo sapiens sapiens* (ho-mo SAY-pee-enz SAY-pee-enz), or "Cro-magnon (kro-MAN-yun) man," seemed to have a utensil for every need. These tools included finely crafted points and blades, needles, harpoons, fishhooks, nets, and snares.

Ancient humans were certainly not the strongest, fastest, or hardiest of animals, but by developing helpful tools they were able to get food, defend themselves, and protect themselves from the harsh environment.

With the making of handy tools and weapons, early humans were able to cook, hunt, and protect themselves.

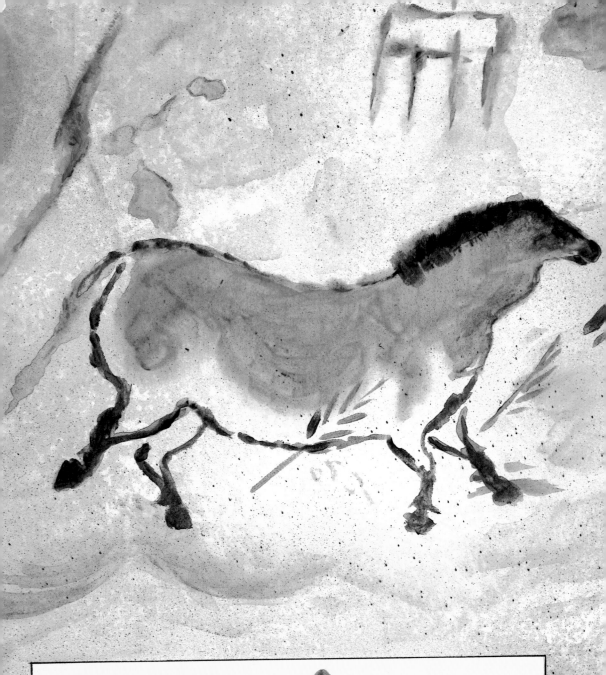

Horse carved in mammoth ivory 30,000 years ago

Woman carved in soapstone 20,000 years ago

Female head carved in ivory 18,000 years ago

Ancient Art

One day in 1940, four young boys were playing in the countryside of southwestern France. They had no idea that before the day was over they would discover one of the world's finest examples of prehistoric art. A small dog trailed along with the boys, barking and chasing after rabbits. After a while, the dog disappeared, and the worried youngsters searched for him. When they found him, they also found the entrance to a huge cave that would soon be known to the world as Grotto de Lascaux (GRAHT-o day las-KO).

The walls and ceilings of the main cavern and of several smaller chambers are covered with primitive paintings and engravings of deer, oxen, and horses. The magnificent paintings were created by humans between 22,000 and 28,000 years ago. Outlined in deep black, the figures are stained in yellow, red, brown, and black. The colors were made of rocks ground into powder and mixed with water or animal fat. Iron and manganese created rich red and black. The dull yellow was crushed iron carbonate. By the light of lamps that burned animal fat for fuel, ancient artists rubbed the colors onto the walls of the cave with animal skin or blew it on dry through hollow bones. The artists stored their precious paints in animal bones and skulls. The animal paintings at Lascaux and other sites may have been made during special ceremonies. Perhaps the artists believed that by drawing a picture of a successful hunt, they would have good luck during a real one.

Cave paintings are not the only examples of ancient art. Thousands of years ago, craftsmen of Eurasia carved figures of humans and animals from ivory, bone, clay, and soapstone. They created jewelry and they decorated their clothing with teeth, feathers, and bone. Made of teeth and shells, the earliest necklaces ever found are at least 35,000 years old!

This painting of a horse from the Grotto de Lascaux and various sculptures tell us about the early humans.

Lessons from the Past

In the opening pages of the book of life on Earth, the highest forms of life were tiny one-celled creatures drifting in an ancient ocean. Millions and millions of years later, dinosaurs ruled the land. The death of the mighty dinosaurs signaled the beginning of the reign of mammals. Now humans have become the dominant life-form on this planet. Will we survive as long or as well as the many successful species that have gone before us?

Thousands of years ago, the fortunes of the rather frail human species depended on its intelligence and ability to reason, solve problems, and adapt to the surroundings. Today we often adapt our surroundings to our own needs. In the last 10,000 years, we have learned to plant and harvest our own food, to mine and use metals, and to record that information in written form. In a geological "instant," humans have progressed from hand axes to lasers. And the pace of learning is picking up. One hundred years ago, people traveled by horse-drawn carriages. There were no televisions, computers, or airplanes. Today we are sending probes to other planets and beyond. But in the race to advance, we have fouled the air and water, destroyed forests, and brought about the early extinction of thousands of plants and animals. Now we must use our intelligence and reason to restore the earth and to protect the plants and animals that share the planet with us. Then we will be prepared to meet the future, proud of our place in the great book of life.

For Further Reading

Arnold, Caroline: *Trapped in Tar: Fossils from the Ice Age,* New York City, Houghton Mifflin Company, 1987.

Berger, Melvin: *The New Earth Book,* New York City, Thomas Y. Crowell, 1980.

Dixon, Dougal; Cox, Barry; Savage, R. J. G.; and Gardiner, Brian: *The Macmillan Illustrated Encyclopedia of Dinosaurs and Prehistoric Animals,* New York City, Macmillan Publishing Co., 1988.

Lambert, David: *The Field Guide to Prehistoric Life,* New York City, Facts on File, 1985.

Lye, Keith: *Our World: Mountains,* Morristown, New Jersey, Silver Burdett Press, 1987.

Raymo, Chet: *The Crust of Our Earth,* New York City, Prentice Hall Press, 1983.

Ricciuti, Edward: *Older Than the Dinosaurs,* New York City, Thomas Y. Crowell, 1980.

Sattler, Helen Roney: *Hominids: A Look Back at Our Ancestors,* New York City, Lothrop, Lee & Shepard Books, 1988.

Stein, Sara: *The Evolution Book,* New York City, Workman Publishing, 1986.

Wilkinson, Phil (ed.): *Eyewitness Books: Early Humans,* New York City, Alfred A. Knopf, 1989.

Index

Africa 11, 27, 53, 57
Amebelodon 45
amphibians 21, 23
   Diadectes 21
   Diplocaulus 21
   Ichthyostega 21
   Platyhystrix 21
Antarctica 27
Arsinoitherium 43
Asia 27, 47, 51
Atlantic Ocean 11, 53
Australopithecus 57

Basilosaurus 35
birds 29, 39
   Aepyornis 29
   Argentavis 29
   Diatryma 29
body temperature 21, 25, 31
Burgess Shale 9

cave lion 37
cave paintings 51, 61
Cenozoic Era 7, 29
   Quaternary Period 29
      Pleistocene Epoch 29, 37, 39, 41, 45, 51
   Tertiary Period 29
      Eocene Epoch 35, 43
      Miocene Epoch 35, 47
      Oligocene Epoch 43
continental drift 53
   mountain building 27
   rifts 11
coral 13
cynobacteria 5

dinosaurs 25, 31, 62
Diprotodon 33
dire wolf 39
Doedicurus 41

Enaliarctos 35
Eohippus 47
Equus 47
Europe 11, 37, 51, 53

Fish 13, 21, 23, 35
   Arandaspis 15
   Dinichthys 15
   lobe-finned fish 15, 21
   sharks 15

geologic time 7
   atomic dating 7
   carbon 14, 7
Gigantopithecus 55
Glyptodon 41

Hallucigenia 9
Himalayas 27, 55
humans 31, 51, 57, 59, 61, 62
Hyracotherium 47

Iapetus 11
Ice Age 45, 51
India 27, 49, 55
Indricotherium 49
insects 9, 19, 23, 31, 39
   beetles 23
   bristletails 19
   cockroaches 23
   Meganeura 23
   springtails 19

mammals 29, 31, 33, 35
   ape 31, 55
   deer 51
   dog 39
   dolphin 35
   duck-billed platypus 33
   echidna 33
   elephant 45
   ground sloth 41
   horse 43, 47
   marsupials 33
   monotremes 33
   pinnipeds 35
   rhinoceros 49
   Smilodon 37
   whale 35
mammoths 45
mastodonts 45
Mediterranean Sea 53
Megaloceros 51
Megatherium 41

Megazostrodon 33
Megistotherium 33
Merychippus 47
Mesohippus 47
Mesozoic Era 7, 21, 31
   Cretaceous Period 33
   Jurassic Period 21
millipede 19

nautiloid 13
North America 11, 37, 45, 47

Pakicetus 35
Paleozoic Era 7, 9, 13, 23, 25
   Cambrian Period 11
   Carboniferous Period 19, 23
   Devonian Period 15, 19, 21
   Ordovician Period 13
   Silurian Period 13, 15
Pangaea 11, 27
Phenacodus 43
Phlaocyon 39
plants 5, 17, 19, 23, 31, 62
Precambrian Era 7
Pterygotus 13
Purgatorius 31

reptiles
   Cynognathus 25
   Dimetrodon 25

Smilodon 37
South America 29, 41, 53
spiders 9, 19

teeth
   fossil 55
   mammals 33, 37, 39, 43, 45, 57, 59
   reptiles 25
Tethys Sea 53
trilobites 9, 13

Uintatherium 43

Walcott, Charles 9